EXPLORE!

FORCE OF NATURE

HURRICANES

Enslow PUBLISHING

BY MONIKA DAVIES

Please visit our website, www.enslow.com. For a free color catalog of all our high-quality books, call toll free 1-800-398-2504 or fax 1-877-980-4454.

Cataloging-in-Publication Data

Names: Davies, Monika.
Title: Hurricanes / Monika Davies.
Description: New York : Enslow Publishing, 2021. | Series: Force of nature | Includes glossary and index.
Identifiers: ISBN 9781978518452 (pbk.) | ISBN 9781978518469 (library bound)
Subjects: LCSH: Hurricanes–Juvenile literature.
Classification: LCC QC944.2 D383 2021 | DDC 551.55'2–dc23

Published in 2021 by
Enslow Publishing
101 West 23rd Street, Suite #240
New York, NY 10011

Designer: Katelyn E. Reynolds
Editor: Monika Davies

Photo credits: Cover, p. 1 FotoKina/Shutterstock.com; cover, pp. 1–48 (series art) Merfin/Shutterstock.com; pp. 5, 20 NOAA via Getty Images; p. 7 Artisticco/Shutterstock.com; p. 8 courtesy of NASA; p. 9 BlueRingMedia/Shutterstock.com; p. 10 udaix/Shutterstock.com; p. 11 Dorling Kindersley/Getty Images; p. 12 Mike Mareen/Shutterstock.com; p. 13 Denis Burdin/Shutterstock.com; pp. 14, 16 Drew McArthur/Shutterstock.com; p. 15 VectorMine/Shutterstock.com; p. 17 Alan Budman/Shutterstock.com; p. 18 Warren Faidley/Corbis/Getty Images Plus; p. 19 CHUYN/E+/Getty Images; p. 21 NASA images/Shutterstock.com; p. 22 NOAA/Saperaud~commonswiki/Wikipedia.org; p. 23 Steven Hunt/Photographer's Choice/Getty Images Plus; p. 24 Zern Liew/Shutterstock.com; p. 25 Placebo365/E+/Getty Images; p. 27 Fabrice Coffrini/AFP via Getty Images; p. 28 Zack Wittman/Bloomberg via Getty Images; p. 29 Kyle Niemi/U.S. Coast Guard via Getty Images; p. 30 Jose Jimenez/Getty Images; p. 31 Captain Cobi/Shutterstock.com; p. 32 Rhona Wise/AFP via Getty Images; p. 33 Jim Watson/AFP via Getty Images; pp. 34, 45 Joe Raedle/Getty Images; p. 35 Tad Denson/Shutterstock.com; p. 37 courtesy of the Library of Congress; p. 38 Robert Sullivan/AFP via Getty Images; p. 39 Steve Starr/Corbis/Corbis via Getty Images; p. 40 ParkerDeen/E+/Getty Images; p. 41 POOL/AFP via Getty Images; p. 42 Chris Seward/Raleigh News & Observer/Tribune News Service via Getty Images; p. 43 Ricardo Arduengo/AFP via Getty Images; p. 44 Jodi Jacobson/E+/Getty Images.

Printed in the United States of America

Some of the images in this book illustrate individuals who are models. The depictions do not imply actual situations or events.

CPSIA compliance information: Batch #BS20ENS: For further information contact Enslow Publishing, New York, New York, at 1-800-542-2595.

CONTENTS

WORDS IN THE GLOSSARY APPEAR IN **BOLD** TYPE THE FIRST TIME THEY ARE USED IN THE TEXT.

CATASTROPHIC CYCLONES

Hurricanes are **violent** tropical storms, known for bringing heavy rainfall, demolishing coastal cities, and causing numerous deaths. Hurricanes destroy buildings, bridges, and vehicles. Their strong winds can also knock down trees and rip apart power lines. These storms are very dangerous.

Every year, an average of six hurricanes form in the Atlantic Ocean, though not all of these will reach land. However, when a hurricane does strike land, considerable **damage** is almost assured. In this book, we'll examine how a hurricane develops, which historic hurricanes left **devastating** aftermaths, and how we can track and prepare for future hurricanes.

EVERY YEAR, HURRICANES CAUSE BILLIONS OF DOLLARS OF PROPERTY DAMAGE. SOME HURRICANES ARE POWERFUL ENOUGH TO DESTROY WHOLE CITIES.

STARTING A HURRICANE

Hurricanes are one of the most powerful and **destructive** forces in nature. In order to understand how these storms form, it's important to first understand the water cycle. It's also key to learn how air pressure and weather systems work together to create hurricanes.

The water cycle begins when the sun's rays warm water on Earth's surface, turning some of it into water vapor. This process is called evaporation. As water vapor rises into the cool upper atmosphere, it turns back into liquid or frozen water. This process, called condensation, is how clouds form. When water drops or ice crystals inside clouds get too heavy, they fall back to Earth as rain, hail, sleet, or snow. This is called precipitation. Then the cycle begins again.

THE WATER CYCLE

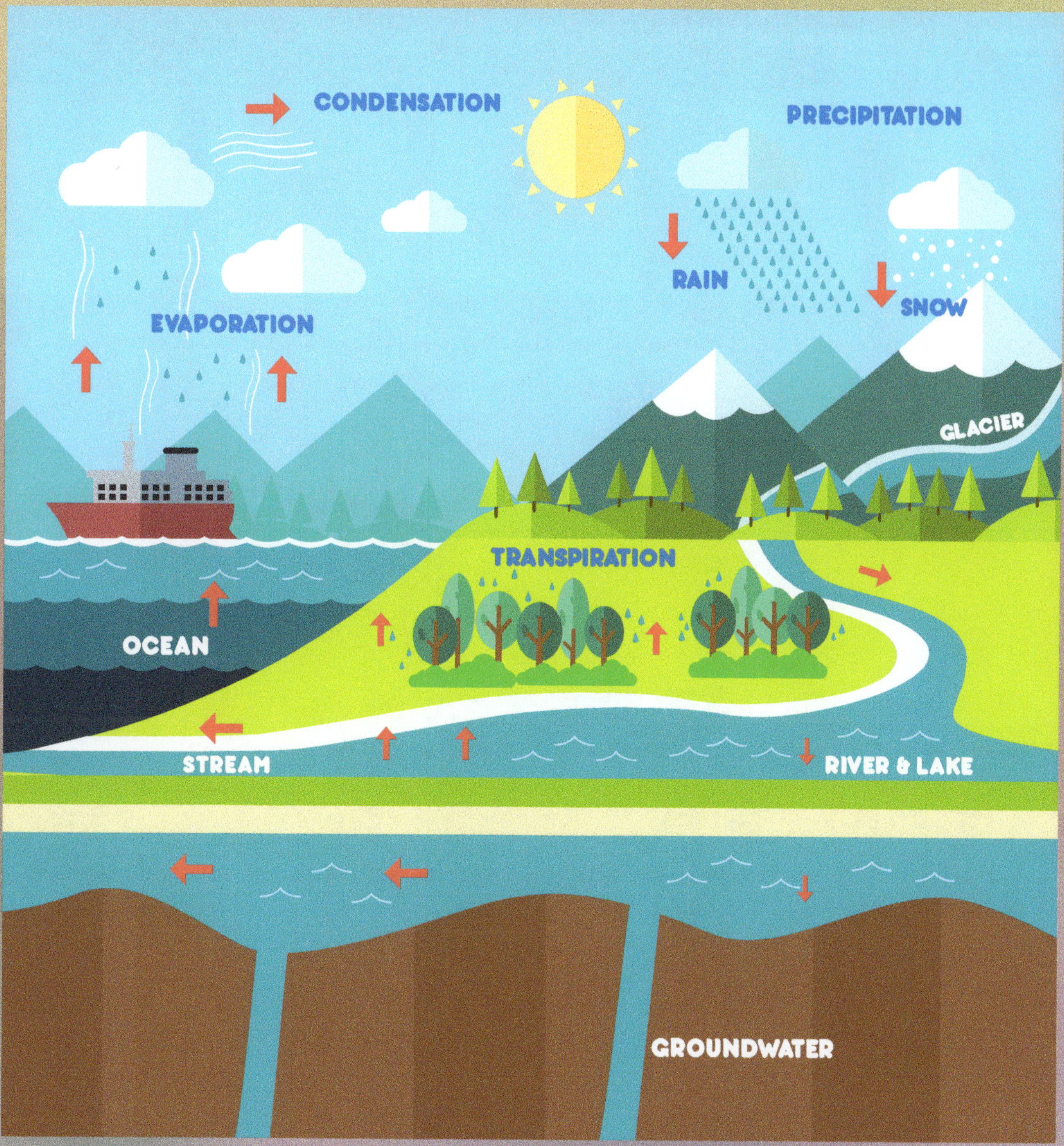

HURRICANES AND THE RAINSTORMS THEY BRING ARE PART OF THE WATER CYCLE. THEY MOVE HUGE AMOUNTS OF WATER FROM ONE PLACE TO ANOTHER.

FUN FACTS

Did you know hurricanes, tropical cyclones, and typhoons are all the same thing? "Hurricane" is the term used for severe storms over the Atlantic Ocean, Gulf of Mexico, Caribbean Sea, and the northeastern Pacific Ocean. In some parts of the Pacific Ocean, hurricanes are called typhoons. In other parts of the Pacific Ocean and in the Indian Ocean, hurricanes are called tropical cyclones. Hurricanes aren't only named differently across the globe. They also spin in different directions! Hurricanes spin counterclockwise above the equator. Below the equator, they spin clockwise. Once a hurricane forms, it's largely steered by the global winds that blow across the oceans. These allow scientists to **predict** a hurricane's path.

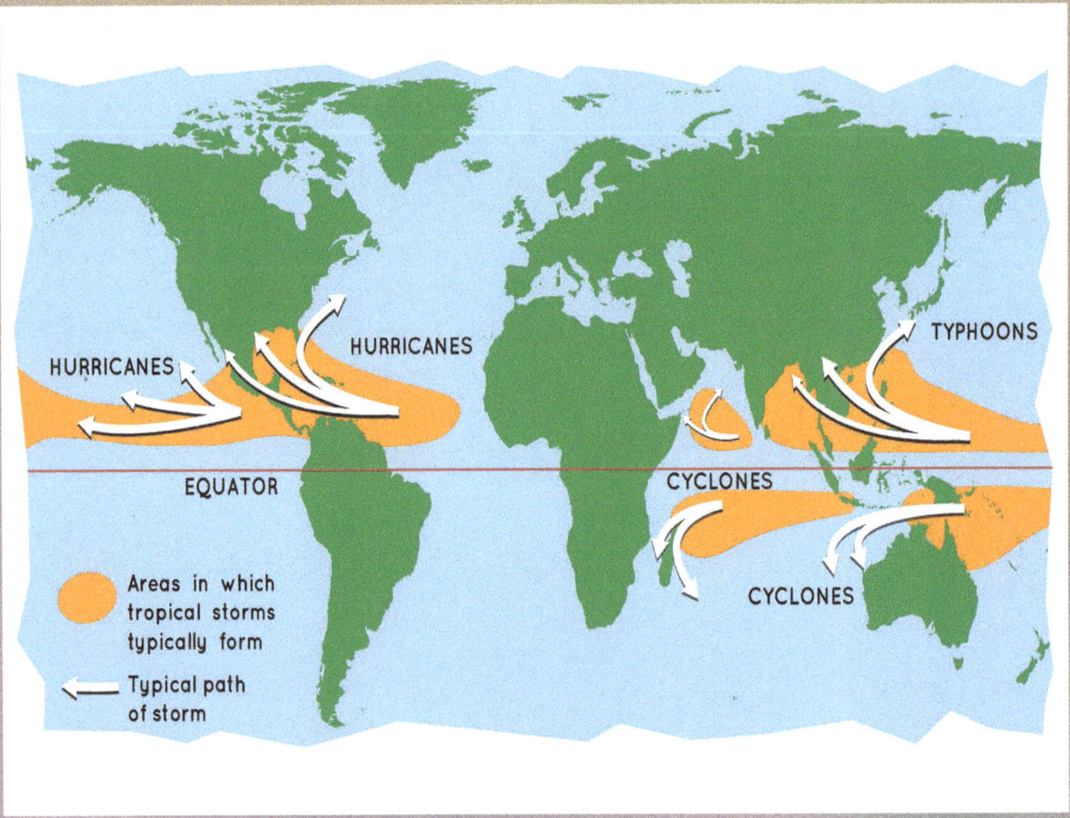

THIS GLOBAL MAP SHOWS THE USUAL PATH OF A STORM DEPENDING ON WHICH WATER BASIN, OR BODY OF WATER, IT ORIGINATES IN.

EQUATOR:
AN IMAGINARY LINE AROUND EARTH THAT IS THE SAME DISTANCE FROM THE NORTH AND SOUTH POLES

EXPLORE MORE

WATER DOESN'T HAVE TO BE ON EARTH'S SURFACE TO EVAPORATE. IT CAN ALSO EVAPORATE FROM PLANTS, ESPECIALLY PLANT LEAVES. THIS PROCESS IS CALLED TRANSPIRATION, AND IT'S ANOTHER PART OF THE WATER CYCLE.

The air we breathe is made up of tiny, **invisible** particles. Even though we don't feel them, air particles have weight, just like everything else on Earth. The weight of air particles as they press down on Earth creates air pressure.

Air pressure depends on how many air particles occupy a space. This amount changes as air temperature changes.

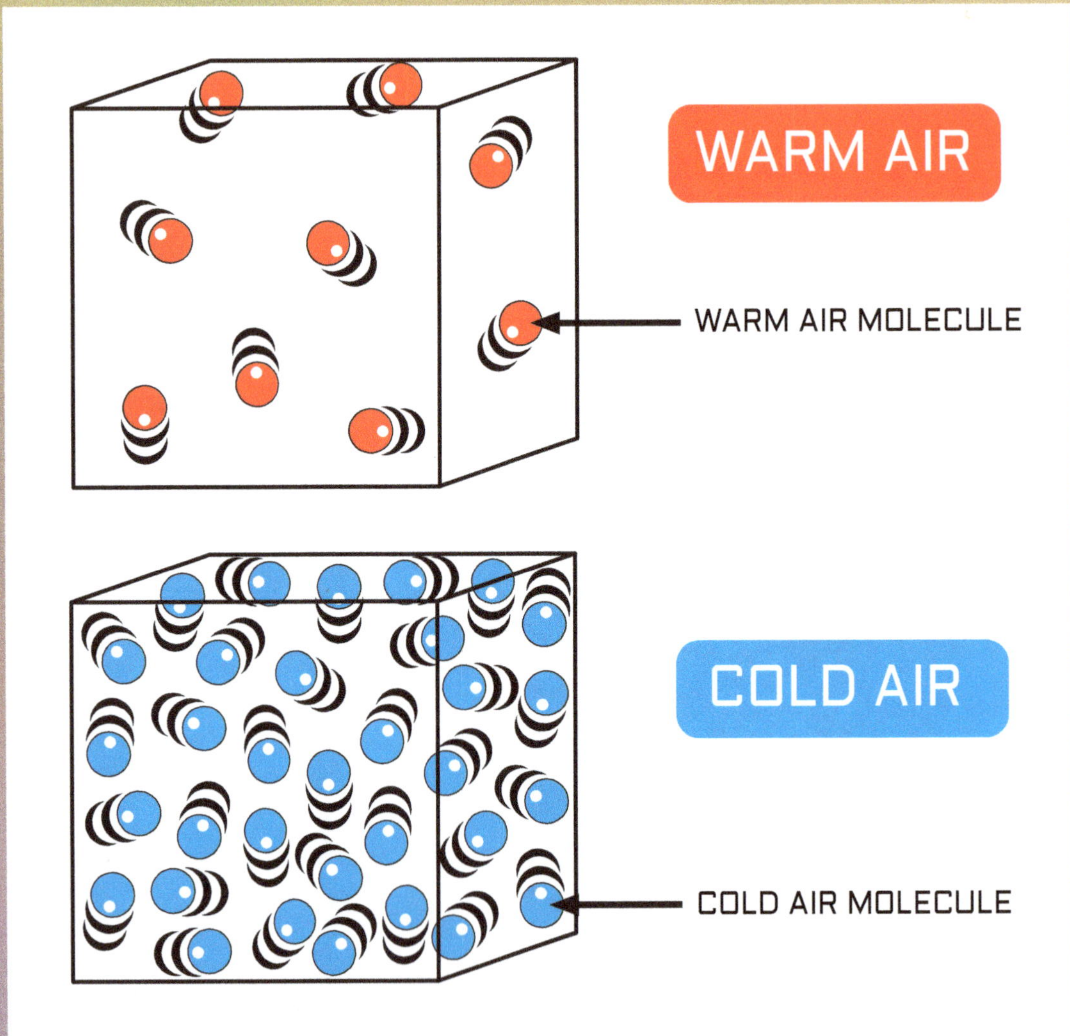

COLD AIR MOLECULES MOVE MORE SLOWLY AND ARE PACKED CLOSELY TOGETHER. WARM AIR MOLECULES ARE MORE ENERGETIC AND MOVE FURTHER APART FROM EACH OTHER.

EXPLORE MORE

COOL AIR IS MORE DENSE, CAUSING IT TO SINK. THIS CREATES HIGH PRESSURE. WARM AIR IS LESS DENSE AND LIGHTER COMPARED TO COOL AIR. THIS CAUSES WARM AIR TO RISE, CREATING LOW PRESSURE.

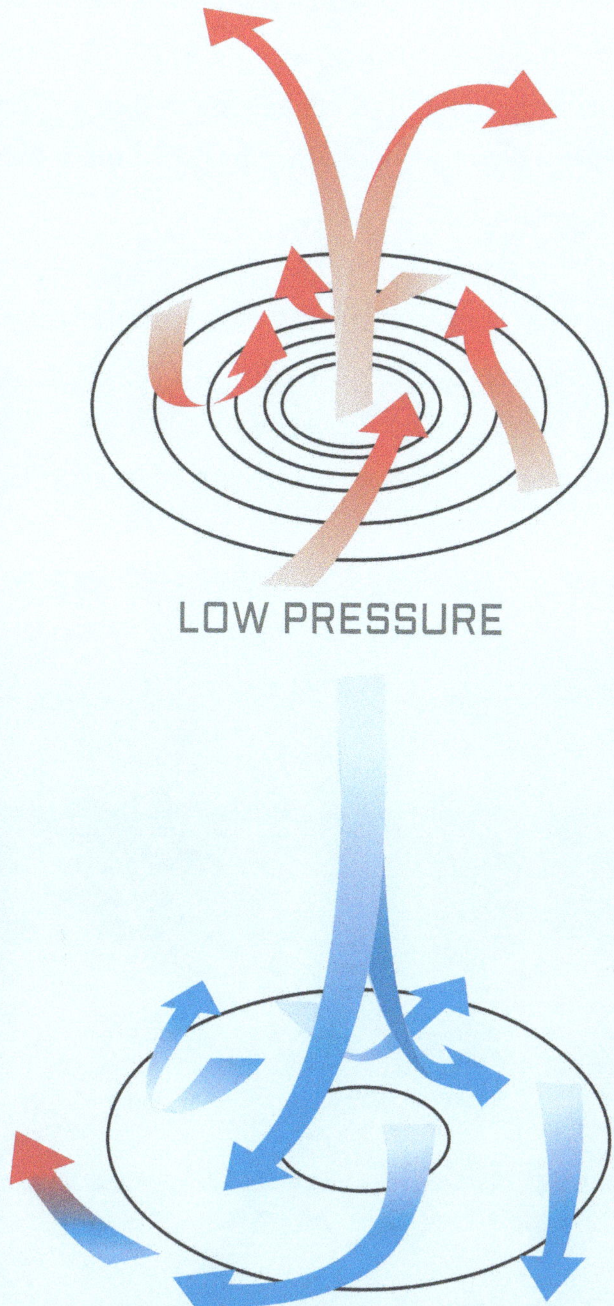

AIR PRESSURE IS THE DOWNWARD FORCE OF AIR CAUSED BY ITS WEIGHT.

One way to predict weather is to measure air pressure. Usually, high-pressure systems mean that weather will be calm and skies will be clear. Low-pressure systems often bring warmer, stormy weather. Hurricanes usually begin their life as a small storm caused by a low-pressure system.

Hurricanes form over ocean waters that are at least 80°F (27°C). As warm, moist air rises, cooler air sinks to take its place. Once that air becomes warm, it also rises. More cool wind rushes in, and wind speed increases as the air cycles. The moist air creates clouds and thunderstorms called tropical disturbances.

HURRICANES NEED WARM, MOIST AIR TO COME TO LIFE. THIS IS WHY HURRICANES FORM ONLY IN TROPICAL WATERS OF A CERTAIN TEMPERATURE.

TROPICAL DISTURBANCE:

A WEATHER EVENT OVER WARM OCEAN WATER THAT LASTS AT LEAST 24 HOURS AND DOESN'T HAVE CONTINUOUS WINDS

THE WORD "TROPICAL" MEANS HAVING TO DO WITH THE WARM PARTS OF EARTH NEAR THE EQUATOR.

A tropical disturbance can grow stronger only if it remains above warm water. A tropical disturbance with wind speeds between 25 and 38 miles (40 to 61 km) per hour is called a tropical depression. If wind speeds reach 39 miles (63 km) per hour, the tropical depression becomes a tropical storm. Once winds reach 74 miles (119 km) per hour, the storm is labeled a hurricane.

A TROPICAL STORM CAN BECOME A HURRICANE IN JUST A FEW HOURS. OTHER TIMES, IT CAN TAKE SEVERAL DAYS FOR A HURRICANE TO FORM. AS LONG AS THE WINDS KEEP SPINNING, A HURRICANE IS A POSSIBILITY.

THE FOUR STAGES

A hurricane needs two ingredients to form: warm, moist air and strong winds. A hurricane develops through four stages. It starts as a tropical disturbance. As the thunderstorm clouds within a tropical disturbance grow larger, the winds in the clouds begin to spin quicker and quicker. The winds spin in a circular pattern. The tropical disturbance then becomes a tropical depression as the winds twist and turn faster and faster. The movement of the winds also then creates a calm center inside the storm, known as the eye. The final stage is when the tropical storm becomes a hurricane. At this point, the storm usually measures around 300 miles (483 km) wide, though this width can vary quite a bit.

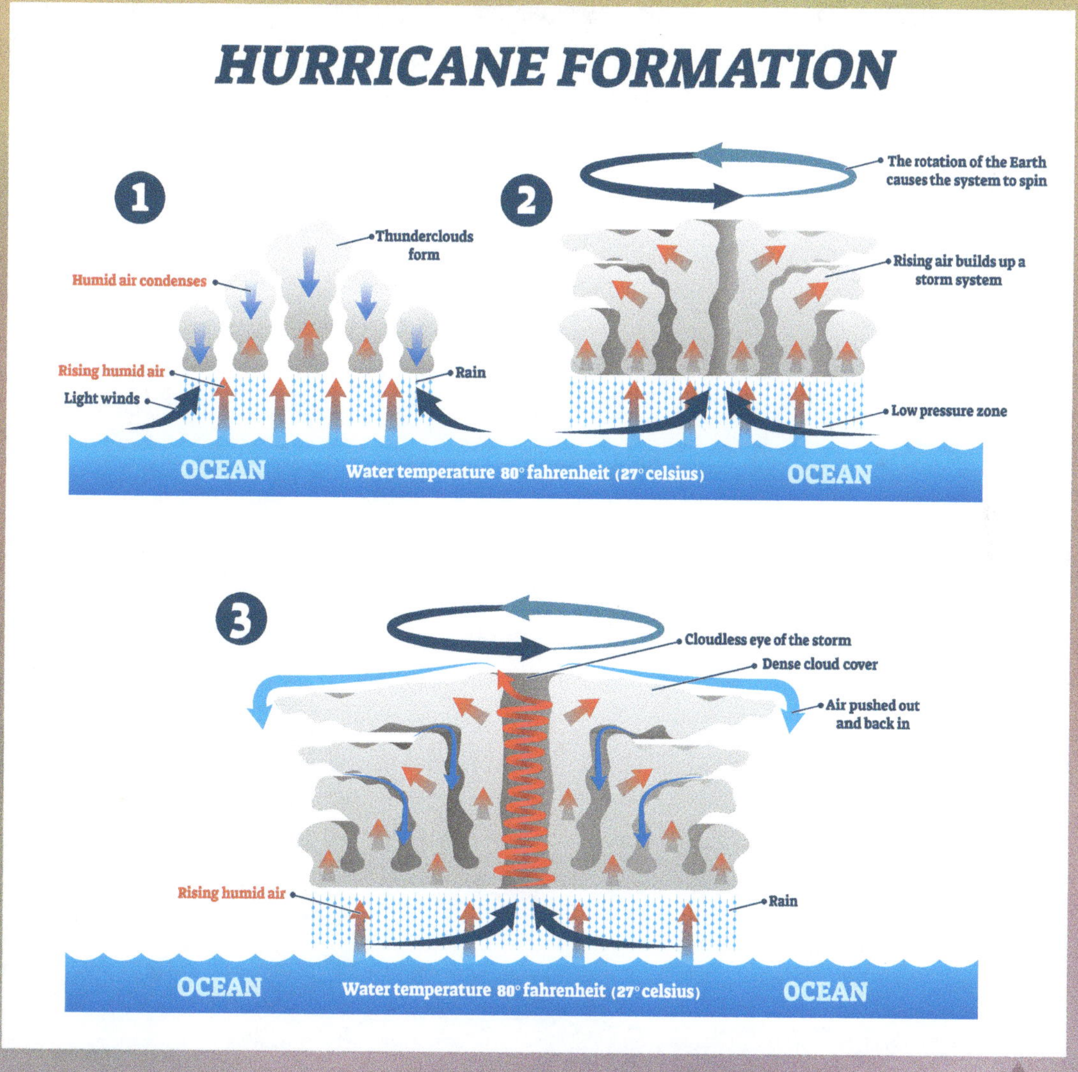

THIS DIAGRAM SHOWS THE FORMATION OF A HURRICANE. NOTICE HOW THE HURRICANE DEVELOPS AS THE WINDS BLOW IN A CIRCULAR MOTION.

HURRICANE FACT FILE

Since hurricanes require warm water and warm air, they usually form only during a certain time of the year. This period is called hurricane season. The official hurricane season lasts from June 1 to November 30. However, hurricanes can occur before or after those dates. In the Atlantic Ocean, the majority of hurricanes occur from August to October.

DURING A THREE-YEAR SPAN OF TIME, THERE ARE GENERALLY ABOUT FIVE HURRICANES THAT MAKE U.S. LANDFALL. USUALLY, TWO OF THOSE HURRICANES ARE "MAJOR" HURRICANES THAT CAUSE CONSIDERABLE DAMAGE AND **CASUALTIES**.

IT'S IMPORTANT TO BE PREPARED WITH SUPPLIES LIKE WATER AND FIRST AID KITS DURING HURRICANE SEASON.

EXPLORE MORE

FORTUNATELY, MOST HURRICANES NEVER MAKE IT TO LAND. SOMETIMES, THE WIND BLOWS THEM BACK OUT TO SEA. OTHER TIMES, A HURRICANE MAY HIT AN AREA OF COLD WATER THAT CAUSES IT TO WEAKEN.

Most hurricanes in the Atlantic Ocean form over a band of warm water that stretches from Africa to Central America. This area is sometimes called Hurricane Alley. Scientists have noticed that the water in Hurricane Alley has been getting warmer. They worry this may cause stronger hurricanes to form.

HURRICANES BEGIN TO WEAKEN AND DIE ONCE THEY HIT LAND. HOWEVER, STRONG WINDS AND HEAVY RAINFALL CAN STILL CAUSE DAMAGE FOR HUNDREDS OF MILES AS THE STORM MOVES ACROSS LAND. A HURRICANE CAN ALSO REGAIN STRENGTH IF IT MOVES BACK OVER WATER AFTER HITTING LAND.

DEADLY DESTRUCTION

Once they hit land, hurricanes cause damage in many ways. Powerful winds can snap trees and power lines, cutting off roads and electricity. Buildings that aren't constructed to withstand heavy winds are often no match for a hurricane. A hurricane can turn ordinary objects such as signs, branches, and pieces of wood into dangerous flying objects. Sometimes, hurricanes can even cause tornadoes to form. As a hurricane reaches shore, its winds can push a large wall of water onto land. This is called a storm surge. Storm surges can cause severe flooding and are often responsible for the most deaths during a hurricane.

FLOOD:
A LARGE STREAM OF RUSHING WATER THAT COVERS LAND THAT'S NORMALLY DRY

The center of a hurricane is called the eye. The eye of a hurricane is usually about 20 to 40 miles (32 to 64 km) wide. Here, air slowly sinks down and grows warmer. Perhaps surprisingly, the low-pressure air in the eye means that it's very calm.

THE EYE OF A HURRICANE IS USUALLY CIRCULAR IN NATURE, BUT IT CAN BE MANY DIFFERENT SIZES AND SHAPES. IN A POWERFUL HURRICANE, THE EYE CAN EASILY BE SPOTTED FROM SPACE.

THE CALM IN A STORM

On land, the eye of the storm is where the hurricane's ferocious winds almost completely die down. The storm's torrential rains come to a halt. If someone were to stand in the eye of a hurricane, they usually won't see many clouds. They may even see the sun! However, **experts** warn against standing or going outside in the eye of a hurricane. Although it may look safe, the dangerous winds and rains of the hurricane are on their way again. The calm in the eye of the storm will last only a brief amount of time.

THIS IMAGE OF THE EYE OF HURRICANE FLORENCE WAS TAKEN FROM THE INTERNATIONAL SPACE STATION!

TORRENTIAL: COMING IN A LARGE, QUICK FLOW OR STREAM

Surrounding the eye is a ring of clouds and thunderstorms called the eye wall. The eye wall is where the strongest winds are found. The eye wall is also the tallest part of a hurricane. Some eye walls can reach 49,000 feet (15,000 m) into the air! How tall is that? The tallest mountain in the world, Mount Everest, is 29,035 feet (8,850 m) tall. This means a hurricane may tower above Mount Everest!

A SPECIAL PLANE TOOK THIS PHOTOGRAPH OF THE EYE OF HURRICANE KATRINA.

EXPLORE MORE

THE EYE WALL IS THE MOST DANGEROUS, OR UNSAFE, SECTION OF A HURRICANE. THIS AREA OF A HURRICANE HAS THE STRONGEST WINDS AND THE HEAVIEST RAINS, BOTH OF WHICH CAUSE THE GREATEST DESTRUCTION.

IN THE NORTHERN HEMISPHERE, THE STRONGEST WINDS IN A HURRICANE ARE OFTEN FOUND ON THE RIGHT-HAND SIDE OF THE STORM.

Rainbands make up the rest of the hurricane. They spiral into the center of the hurricane. They can range from a few miles to tens of miles wide and from 50 to 300 miles (80 to 483 km) long. Rainbands bring heavy rain and winds.

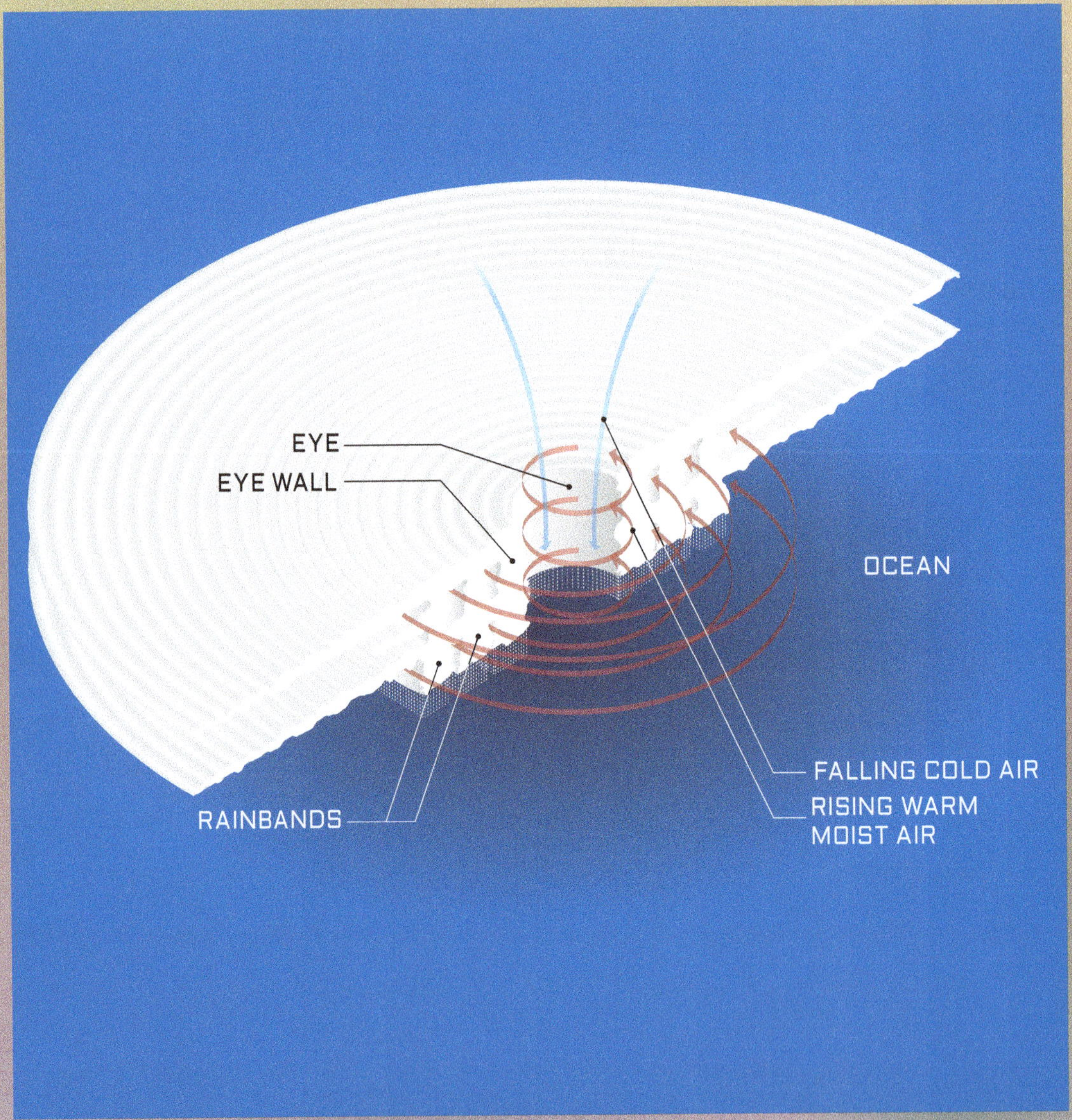

THIS CROSS-SECTION OF A HURRICANE SHOWS WHAT A FULLY FORMED STORM WOULD LOOK LIKE IF YOU COULD GET AN INSIDE LOOK!

RAINBANDS CAN ALSO BRING TORNADOES. THE MAJORITY OF TORNADOES THAT RESULT FROM A HURRICANE ARE CREATED WITHIN THE OUTER, OR OUTSIDE, RAINBANDS. HOWEVER, TORNADOES RESULTING FROM HURRICANE ACTIVITY ARE USUALLY WEAKER THAN TORNADOES THAT BEGIN IN CENTRAL U.S. STATES.

TORNADO:
A COLUMN OF FIERCELY ROTATING WINDS THAT EXTENDS FROM A STORM CLOUD TO THE GROUND

NAMING AND TRACKING HURRICANES

It's common for several hurricanes or tropical storms to occur in different parts of the ocean at the same time. To avoid confusion, every tropical storm is given a name. If a tropical storm becomes a hurricane, it keeps its name. Over the years, names were chosen in many different ways. Today, **forecasters** take names in order from a list.

Hurricane names are chosen from several different lists. Different parts of the world use different lists.

CLIMATE:
THE AVERAGE WEATHER CONDITIONS OF A PLACE OVER A PERIOD OF TIME

EXPLORE MORE

THE WORLD METEOROLOGICAL ORGANIZATION ESTABLISHED THE WAY HURRICANES ARE NAMED. FORECASTERS FOUND THAT SHORT, MEMORABLE NAMES WERE EASIER TO USE IN WRITTEN AND SPOKEN COMMUNICATIONS THROUGHOUT THE WORLD.

THE WORLD METEOROLOGICAL ORGANIZATION IS PART OF THE UNITED NATIONS, WHICH IS A GROUP OF NATIONS THAT UNITED AFTER WORLD WAR II WITH THE INTENTION OF RESOLVING CONFLICTS BETWEEN NATIONS. ONE OF THE ORGANIZATION'S PURPOSES IS TO "PROVIDE VITAL WEATHER AND CLIMATE INFORMATION WORLDWIDE."

The Atlantic Ocean list is different from the one used for the Pacific. On the Atlantic list, the names are listed alphabetically and switch between male and female. There are no names that begin with the letters Q, U, X, Y, or Z. In 2019, some of the Atlantic hurricanes were named Andrea, Dorian, Fernand, Humberto, Karen, Olga, Sebastien, and Wendy.

THIS PHOTOGRAPH SHOWS THE BOARDED-UP WINDOW OF A HOUSE IN COCOA BEACH, FLORIDA. THE BOARD HAS A LIST OF PREVIOUS HURRICANES THAT HIT FLORIDA, INCLUDING HURRICANE MATTHEW IN 2016, HURRICANE IRMA IN 2017, AND HURRICANE DORIAN IN 2019.

HURRICANE KATRINA WAS A PARTICULARLY DEADLY STORM THAT STRUCK NEW ORLEANS IN 2005. IT WASN'T UNTIL 2018 THAT THE CITY PARTIALLY REBUILT ITS HURRICANE DEFENSES. THE TOTAL COST FOR THESE REPAIRS WAS $14.6 BILLION.

EXPLORE MORE

IF A HURRICANE IS PARTICULARLY DEVASTATING AND DEADLY, THEN FORECASTERS RETIRE THE NAME OF THE HURRICANE. THIS MEANS THE NAME WON'T BE USED FOR A FUTURE HURRICANE. FOR EXAMPLE, THE NAME "KATRINA" WAS RETIRED IN 2005 DUE TO HURRICANE KATRINA'S DEADLY DESTRUCTION.

Meteorologists also use a scale to show how powerful a hurricane is. This scale is called the Saffir-Simpson Scale, and it's divided into five categories. A Category 1 hurricane is the least destructive, while a Category 5 is the most destructive.

METEOROLOGIST:
SOMEONE WHO STUDIES WEATHER, CLIMATE, AND THE ATMOSPHERE

HURRICANE DORIAN, A CATEGORY 5 HURRICANE, SWEPT THROUGH THE BAHAMAS IN SEPTEMBER 2019.

CLASSIFYING HURRICANES

The Saffir-Simpson Scale is a system to classify the category, or class, of a hurricane based on the storm's highest sustained wind speed. Winds must be moving at a certain speed for at least one minute to measure as a maximum sustained wind speed. The scale is helpful for guessing the possible scale of property damage of a hurricane. A Category 3 hurricane may cause many trees to be pulled up from their roots, as well as cause damage to roofs of well-built houses. A Category 5 hurricane may be **catastrophic** and lead to the destruction of a large number of framed houses, as well as long-term power outages.

SAFFIR-SIMPSON SCALE

1 SOME DAMAGE

74–95 MILES (119–153 KM) PER HOUR WINDS

2 EXTENSIVE DAMAGE

96–110 MILES (154–177 KM) PER HOUR WINDS

3 DEVASTATING DAMAGE

111–129 MILES (178–208 KM) PER HOUR WINDS

4 CATASTROPHIC DAMAGE

130–156 MILES (209–251 KM) PER HOUR WINDS

5 CATASTROPHIC DAMAGE

157 MILES (252 KM) PER HOUR WINDS OR HIGHER

WHILE CATEGORY 1 AND 2 HURRICANES CAN CAUSE DAMAGE, CATEGORY 3, 4, AND 5 HURRICANES ARE CONSIDERED MAJOR HURRICANES. A MAJOR HURRICANE USUALLY RESULTS IN CONSIDERABLE DESTRUCTION, AS WELL AS A NUMBER OF CASUALTIES.

Since the 1960s, satellites have been used to track hurricanes. But satellites don't always tell forecasters enough to categorize a hurricane. To get more exact information, U.S. Air Force Reserve pilots fly special planes directly into hurricanes. These planes collect **data**, such as wind speed and air pressure. A plane may make several trips through a hurricane.

The National Oceanic and Atmospheric Administration (NOAA) uses special jets to fly around and above hurricanes. The planes' instruments help forecasters predict a hurricane's path by measuring the winds that guide the hurricane.

NOAA SPECIAL JETS SOMETIMES HAVE FUN NICKNAMES! THIS NOAA JET IS KNOWN AS "KERMIT," WHILE ANOTHER IS KNOWN AS "MISS PIGGY."

SATELLITE:
AN OBJECT THAT CIRCLES EARTH IN ORDER TO COLLECT AND SEND INFORMATION OR AID IN COMMUNICATION

U.S. AIR FORCE RESERVE PILOTS WHO FLY INTO HURRICANES ARE KNOWN AS HURRICANE HUNTERS. TWO PILOTS DIRECT AND DRIVE THE AIRCRAFT, WHICH GO RIGHT INTO THE HEART OF A HURRICANE IN SEARCH OF ATMOSPHERIC DATA.

The U.S. military began flying planes through hurricanes in 1944. That year, planes from the navy and the army tracked a hurricane as it moved from Puerto Rico to the northeastern United States. The information that resulted allowed people to prepare for the hurricane and saved many lives.

IT'S IMPORTANT TO ALSO STAY SAFE AFTER A HURRICANE STRIKES. MAKE SURE TO AVOID WADING THROUGH FLOODWATER, AS THE WATERS MAY BE CONTAMINATED, OR POLLUTED. DON'T USE WET ELECTRICAL DEVICES, AND USE FLASHLIGHTS INSTEAD OF CANDLES IF THERE'S NO POWER AT YOUR HOUSE.

PREPARING FOR A HURRICANE

If you live in an area where hurricanes are known to strike, it's important to be prepared. Be sure to keep emergency supplies of food, water, and rain gear in your home. Since hurricanes often cause power loss, it's also a good idea to have a flashlight, a battery-powered radio, and a first aid kit. Always listen to authorities before, during, and after a hurricane. If you're told to leave an area, be sure to do so quickly. Return to your home only after you've been told it's safe. The more you know about hurricanes, the better prepared you will be if one occurs near you.

IF A STRONG HURRICANE IS CLOSE, SOMETIMES GOVERNMENT LEADERS WILL TELL CITIZENS TO EVACUATE, OR LEAVE THE LIKELY DANGEROUS SITUATION. THIS CAN SAVE LIVES!

HURRICANES THROUGHOUT HISTORY

The United States has been hit by many powerful hurricanes over the years. Many hit the East and Gulf Coasts, but these storms can travel many miles inland and continue to cause destruction.

One of the most damaging was Hurricane Camille. In August 1969, this hurricane struck the coast of Mississippi as a Category 5 event. Its winds were so strong they destroyed all the instruments used to measure wind speed. However, based on other information, experts think the wind speeds reached about 200 miles (322 km) per hour.

DISASTER:
AN EVENT THAT CAUSES SERIOUS LOSS OR DESTRUCTION

EXPLORE MORE

ON SEPTEMBER 8, 1900, A POWERFUL HURRICANE STRUCK GALVESTON, TEXAS. THE ENTIRE CITY WAS FLOODED, AND MOST OF IT WAS DESTROYED. IT'S THOUGHT THAT MORE THAN 6,000 PEOPLE WERE KILLED.

TO THIS DAY, THE 1900 GALVESTON HURRICANE IS THE DEADLIEST NATURAL DISASTER IN U.S. HISTORY. THE DEVASTATION DROVE THE U.S. GOVERNMENT TO PUT A GREATER FOCUS ON HOW TO PREDICT A HURRICANE TO PREVENT FUTURE DISASTERS.

In August 1992, Hurricane Andrew ripped across south Florida. The hurricane's winds destroyed much in the hurricane's path, including wind-speed instruments. After passing over Florida, Andrew continued across the Gulf of Mexico until it hit the Louisiana coast.

THESE BOATS WERE WRECKED BY THE STRONG WINDS AND RAINS OF HURRICANE ANDREW IN 1992.

IT'S ESTIMATED THE TOTAL COST OF DESTRUCTION FROM HURRICANE ANDREW WAS AROUND $26.5 BILLION. FOLLOWING HURRICANE ANDREW, SOUTH FLORIDA CHANGED ITS BUILDING CODES. NOW ALL HOMES BUILT IN SOUTH FLORIDA MUST HAVE STORM SHUTTERS OR WINDOWS WITH IMPACT-RESISTANT GLASS.

EXPLORE MORE

AT FIRST, HURRICANE ANDREW WAS LABELED A CATEGORY 4 HURRICANE, BUT IT WAS LATER UPGRADED TO A CATEGORY 5 HURRICANE IN 2002. THE HURRICANE'S SUSTAINED WIND SPEED MEASURED AT 165 MILES (266 KM) PER HOUR WHEN IT HIT SOUTHEASTERN FLORIDA.

When Hurricane Katrina hit Florida on August 25, 2005, it wasn't a very powerful hurricane. However, once it passed through Florida and entered the Gulf of Mexico, it began to strengthen. On August 28, Katrina turned into a Category 5 hurricane. By the time it hit Louisiana on August 29, it had dropped to Category 3 but was still packing a powerful punch. Katrina then continued its destructive path through Louisiana and into Mississippi before it weakened and died.

Florida, Louisiana, Mississippi, and Alabama all suffered severe wind damage from Hurricane Katrina. Storm surges caused flooding along the coasts of Mississippi and Louisiana.

IT TOOK SOME COMMUNITIES YEARS TO RECOVER AFTER HURRICANE KATRINA. SOME COMMUNITIES NEVER RECOVERED.

MOST OF THE CITY OF NEW ORLEANS, LOUISIANA, WAS FLOODED, WITH 20 FEET (6 M) OF WATER COVERING SOME AREAS. HURRICANE KATRINA WAS THE MOST EXPENSIVE STORM IN U.S. HISTORY. IT CAUSED OVER $75 BILLION IN DAMAGE.

In August 2011, Hurricane Irene created a wide path of destruction from North Carolina to Vermont. Irene's rains caused major flooding. Roads and bridges were carried away by water. Hundreds of thousands of people had to evacuate their homes.

EVACUATE:
TO WITHDRAW FROM A PLACE FOR PROTECTION

HURRICANE IRENE WAS SO STRONG AND DROPPED SO MUCH RAIN THAT IT WASHED AWAY ROADS, LIKE THIS ONE IN NORTH CAROLINA.

MORE HISTORIC HURRICANES

In October 2012, Hurricane Sandy caused severe destruction, totaling around $65 billion in damage. The hurricane stormed along a lengthy path, hitting Haiti, Jamaica, Cuba, and the coasts of New Jersey and New York. It led to at least 147 deaths and was nicknamed "Superstorm Sandy" because of how wide-ranging the storm was. In August 2017, Hurricane Harvey reached the Texas coastline, resulting in severe flooding. The Category 4 hurricane had a record-setting rainfall, with 60.58 inches (153.87 cm) of rainfall recorded in Nederland, Texas. At least 68 people died directly from the storm.

IN SEPTEMBER 2017, HURRICANE MARIA STORMED THROUGH THE ISLAND OF DOMINICA AND THE U.S. VIRGIN ISLANDS, AND LATER HIT PUERTO RICO. THE HURRICANE WAS INCREDIBLY DEADLY AND DESTRUCTIVE, KILLING AN ESTIMATED 2,975 PEOPLE IN PUERTO RICO ALONE.

Over the years, hurricanes have caused mass destruction in the United States. However, as technology develops, it has become easier to predict the occurrence and paths of severe hurricanes. While hurricanes are a deadly force of nature, there are many ways people can prepare for these violent storms.

PLACES WHERE HURRICANES ARE COMMON OFTEN HAVE PLANS IN PLACE FOR HURRICANE CLEAN UP. HOWEVER, IT'S OFTEN NOT ENOUGH.

THE EFFECTS OF CLIMATE CHANGE

Climate change has led to global temperatures increasing, particularly ocean surface temperatures, as well as higher sea levels. There is no long-term evidence that climate change will lead to more hurricanes. However, scientists believe climate change is likely to lead to stronger and more intense hurricanes. As ocean surface temperatures rise, this will likely lead to stronger wind speeds in tropical storms. And, as sea levels rise, this will likely lead to stronger storm surges during a hurricane. Both elements would heighten the impact and devastation of a hurricane.

IT CAN TAKE MANY YEARS FOR A CITY TO CLEAN UP DEBRIS AND REBUILD BUILDINGS FOLLOWING THE DEVASTATION OF A HURRICANE. REBUILDING REQUIRES GOVERNMENT FUNDING AND ASSISTANCE. THIS PHOTOGRAPH SHOWS THE 2018 RECOVERY EFFORTS IN PANAMA CITY, FLORIDA, FOLLOWING HURRICANE MICHAEL.

CLIMATE CHANGE: LONG-TERM CHANGE IN EARTH'S CLIMATE, CAUSED PARTLY BY HUMAN ACTIVITIES SUCH AS BURNING OIL AND NATURAL GAS

GLOSSARY

casualty A person who gets hurt or dies in an event.

catastrophic Causing sudden and considerable damage.

damage Harm; also, to cause harm.

data Facts and figures.

destructive Causing damage or ruin.

devastating Causing widespread damage.

expert Someone who knows a great deal about something.

forecaster Someone who makes informed guesses about future weather.

invisible Unable to be seen.

predict To guess what will happen in the future based on facts or knowledge.

violent Having to do with the use of force to harm someone or something.

FOR MORE INFORMATION

BOOKS

Challoner, Jack. *Eyewitness: Hurricane & Tornado.* New York, NY: DK Publishing, 2014.

Kostigen, Thomas. *Extreme Weather: Surviving Tornadoes, Sandstorms, Hailstorms, Blizzards, Hurricanes, and More!* Washington, DC: National Geographic, 2014.

Patrick, Denise Lewis. *Finding Someplace.* New York, NY: Henry Holt & Co., 2015.

Philbrick, Rodman. *Zane and the Hurricane: A Story of Katrina.* New York, NY: The Blue Sky Press, 2014.

WEBSITES

How Does a Hurricane Form?
scijinks.gov/hurricane
Discover how a tropical cyclone goes from a tropical disturbance to a hurricane.

Hurricanes
www.ready.gov/hurricanes
Learn more about how to be prepared if a hurricane strikes.

Hurricanes Impact Our Lives
eo.ucar.edu/kids/dangerwx/hurricane1.htm
Find out more about the impact of hurricanes.

Publisher's note to educators and parents: Our editors have carefully reviewed these websites to ensure that they are suitable for students. Many websites change frequently, however, and we cannot guarantee that a site's future contents will continue to meet our high standards of quality and educational value. Be advised that students should be closely supervised whenever they access the internet.

INDEX